Cambridge Elements ≡

Elements in Geochemical Tracers in Earth System Science
edited by
Timothy Lyons
University of California
Alexandra Turchyn
University of Cambridge
Chris Reinhard
Georgia Institute of Technology

THE PYRITE TRACE ELEMENT PALEO-OCEAN CHEMISTRY PROXY

Daniel D. Gregory
University of Toronto

T0224854

CAMBRIDGE
UNIVERSITY PRESS

CAMBRIDGE
UNIVERSITY PRESS

University Printing House, Cambridge CB2 8BS, United Kingdom

One Liberty Plaza, 20th Floor, New York, NY 10006, USA

477 Williamstown Road, Port Melbourne, VIC 3207, Australia

314–321, 3rd Floor, Plot 3, Splendor Forum, Jasola District Centre, New Delhi – 110025, India

79 Anson Road, #06–04/06, Singapore 079906

Cambridge University Press is part of the University of Cambridge.

It furthers the University's mission by disseminating knowledge in the pursuit of education, learning, and research at the highest international levels of excellence.

www.cambridge.org
Information on this title: www.cambridge.org/9781108810524
DOI: 10.1017/9781108846974

© Daniel D. Gregory 2020

First published 2020

A catalogue record for this publication is available from the British Library.

ISBN 978-1-108-81052-4 Paperback
ISSN 2515-7027 (online)
ISSN 2515-6454 (print)

The Pyrite Trace Element Paleo-Ocean Chemistry Proxy

Elements in Geochemical Tracers in Earth System Science

DOI: 10.1017/9781108846974
First published online: November 2020

Daniel D. Gregory
University of Toronto

Author for correspondence: Daniel D. Gregory, dgregory@es.utoronto.ca

Abstract: The use of the trace element content of sedimentary pyrite as a proxy for the trace element composition of past oceans has recently emerged. The pyrite proxy has several potential advantages over bulk sample analysis: preservation through metamorphism; little dilution during analysis (samples are ablated not dissolved, allowing for the less abundant elements commonly held in the sulfide fraction to be investigated as proxies); accurate measurement of several elements simultaneously; the ability to screen sediments for hydrothermal overprint; and the technique can give information regarding trace element availability at multiple stages of diagenesis. Because of these multiple strengths, the pyrite trace element proxy is a valuable potential addition to the paleo-ocean chemistry tool kit.

Keywords: LA-ICPMS, pyrite, paleoredox, chalcophile, ocean chemistry

ISBNs: 9781108810524 (PB), 9781108846974 (OC)
ISSNs: 2515–7027 (online), 2515–6454 (print)

Contents

1 Introduction

Pyrite is a common mineral that forms in the water column or sediments when sulfidic conditions are present. During formation, pyrite can incorporate a number of different trace elements (Gregory et al., 2015a), with chalcophile and siderophile trace elements being the most common. In this context, 'incorporate' means both substituted for major elements into the crystal structure as well as held as inclusions of different mineral species within the pyrite. Importantly, the trace element content of pyrite scales with the trace element content of the water from which the pyrite formed (Gregory et al., 2014), albeit in a manner that is not currently well understood. Thus, the trace element content of pyrite provides a unique opportunity to capture information regarding the trace element content of the ocean or sediment pore fluids. This is of particular value because the trace element content of pyrite appears to be unaltered up to mid-greenschist facies metamorphism (Large et al., 2007; Gregory et al., 2019a), which could allow glimpses of the chemistry of the oceans from sedimentary basins that are too metamorphosed for traditional bulk sample proxies to be effective. Furthermore, when coupled with detailed textural analysis, the technique can be applied to understand not only water column chemistry, but also the pore water chemistry during diagenesis and thus the availability of different trace elements to organisms at different stages of diagenesis (Gregory et al., 2019b).

A common critique faced by paleo-ocean chemistry studies is whether the metal enrichments measured are due to changes in ocean chemistry or inputs from distal hydrothermal fluids. Because there are abundant published data of hydrothermal pyrite from different types of ore deposit (Genna and Gaboury, 2015; Belousov et al., 2016; Cook et al., 2016; Gadd, Layton-Matthews et al., 2016; Mukherjee and Large, 2017; Keith et al., 2018; Tardani et al., 2017; Gadd, Peter et al., 2019; Román et al., 2019), pyrite chemistry can be used to confirm or refute interpretations that a given sedimentary rock has been overprinted by hydrothermal fluids (Gregory et al., 2017; 2019a).

1.1 Past Work

Attempts have been made for some time to utilize the trace element content of pyrite to understand depositional settings, overlying water column chemistry, and diagenetic processes. Initially sequential extraction techniques were utilized (Huerta-Diaz and Morse, 1990; 1992). While informative, the use of nitric acid to dissolve the sulfide minerals indiscriminately dissolves all sulfide minerals present and potentially different generations of pyrite. Furthermore, it also mobilizes trace elements held by organic matter and thus may not correctly represent the trace element content of the pyrite of interest. Others have isolated pyrite from sedimentary rocks using a combination of crushing, gravity separation, and hand picking prior to geochemical analysis (Berner et al., 2013; Pisarzowska et al., 2014). While this methodology can successfully select only pyrite to be analyzed, it is a very time-consuming method. Furthermore, it has the limitation that different textures of pyrite still cannot be easily separated and very-fine-grained pyrite is still difficult to extract and may be lost.

In the past five to ten years, great advancements have been made in the application of laser ablation inductively coupled plasma mass spectrometry (LA-ICPMS) to sulfide analysis. This procedure involves the use of a small (10–100 μm) laser beam to vaporize an area of interest, which is then analyzed by a mass spectrometer (details in Section 3.2). The application of this technique allows for the selection of different textures of pyrite; thus, different generations of pyrite can be individually targeted based on the research aims. Since the LA-ICPMS technique also has very low detection limits (ppb levels) and allows for > 30 elements to be analyzed at a time, it provides the researcher an extensive suite of high quality data to consider as an indication of water chemistry at the time of formation and compare with other geochemical proxies.

2 Systematics of the Pyrite Trace Element Proxy

2.1 Proof of Concept: Compilation through Geologic Time

The first study to utilize pyrite trace element content to try and understand past ocean and atmospheric chemistry was presented by Large et al., (2014) who compiled a database of pyrite trace element content from a variety of different sedimentary formations through geologic time. These data provided a strong first-order correlation with existing whole rock studies (Scott et al., 2008; Konhauser et al., 2009; Fig. 1) that suggested that the pyrite trace element content could also be used to understand past atmospheric and ocean chemistry. Large et al. (2014) used five key points to support the use of pyrite chemistry as a proxy of ocean and atmospheric chemistry.

1 The high amounts (60–85 percent) of pore fluid in marine sediments (Baldwin and Butler, 1985; Harrold et al., 1999) means that even when formed within sediments, pyrite will be forming under conditions strongly related to ocean water. Huerta-Diaz and Morse (1992) showed that in anoxic and euxinic sediments, trace element content can vary from 1.4 to 6 times (for Co, Mo, As, and Ni) down core (i.e., with time); however, this is far less than the 2 to 4 orders-of-magnitude changes (for Co, Mo, As, Ni, and Se) that

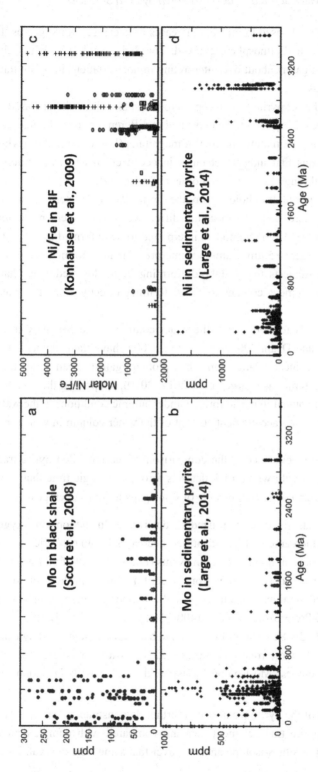

Figure 1 Comparison between compilation of pyrite trace element chemistry (b and d for Mo and Ni respectively) with bulk rock Mo in shales (a) and Ni in banded iron formations (c). Note the correlation between enrichments in bulk sample analysis and pyrite analysis from the same time periods. Modified after Large et al., (2014).

Large et al. (2014) used to indicate increases or decreases in oxygen as the Earth's oceans and atmosphere evolved. That being said, this should also caution investigators about overinterpreting minor variations in pyrite trace element trends.

2 Previous studies of pyrite from Neoproterozoic shales showed that, based on texture, pyrites interpreted to be deposited at different periods of diagenesis tend to vary approximately an order of magnitude (Large et al., 2007) while pyrite variations that indicate changes in ocean chemistry, as mentioned before, vary through time by 2 to 4 orders of magnitude.

3 Analyses of pyrite framboids from the Perth Basin, Western Australia, showed that there was no substantial difference in trace element content between small (≤ 5 μm) framboids interpreted to have formed in the water column and larger (> 5 μm) framboids interpreted to have formed in the pore waters. This suggests that framboids forming in shallow sediments have similar trace element content to those forming directly from the water column.

4 Pyrite framboids formed within the water column in the euxinic Cariaco Basin (Piper and Dean, 2002; Lyons et al., 2003) have trace elements that approximate a linear relationship when plotted against ocean water trace element concentrations (Large et al., 2014; 2019), supporting the idea that trace element content in pyrite formed under euxinic conditions in the water column reflect the trace element content of the water column in which they formed.

5 The fifth line of evidence is the comparison of the compiled pyrite trace element dataset with whole rock datasets through geologic time that show strong first order correlation in spite of some gaps in time coverage (Fig. 1).

The technique has since been utilized to determine fluctuations in oxygen through the Phanerozoic and has identified decreases in atmospheric oxygen coincident with several mass-extinction events (Large et al., 2015; Large et al., 2019). Similar paleo-ocean studies from select timeframes in the Proterozoic have highlighted significant fluctuations and a temporary increase in oxygen during the mid-Proterozoic, around 1400 Ma (Mukherjee and Large, 2016; Mukherjee et al., 2018a). These findings have been further supported by using a combination of whole rock techniques (redox sensitive trace elements, Fe speciation, S-isotopes, Mo-isotopes) (Diamond et al., 2018; Diamond and Lyons, 2018).

The strength of the pyrite technique is further demonstrated in specific case studies when pyrite trace element data are compared with complementary whole rock and in-situ S-isotope analyses over that same time interval. Using a combination of LA-ICPMS pyrite trace element (As, Se, and weak Mo), SHRIMP S-isotope (strong negative $\delta^{34}S$ shift), and whole rock redox sensitive trace element analyses (weak U and Mo in interval identified by Anbar (2007)), Gregory et al. (2015b) suggested there were early pulses of oxygen prior to the Great Oxygenation Event (GOE) during the deposition of the McRea Shale and, more cryptically, the Bee Gorge Member, Western Australia. Similarly, Gregory et al. (2017) used the pyrite technique to support the identification of pulses of oxygen prior to the Neoproterozoic Oxygenation Event (NOE) during deposition of the Doushantuo Formation (see Section 4.1).

More recently, Large et al. (2019) used the correlation between Se/Co ratio in sedimentary pyrite and atmospheric oxygen (as derived from the models of Berner et al., (2006) and inclusions in halite (Blamey et al., 2016)) to model past oxygen concentration. They used this model to argue that atmospheric oxygen was much higher in the middle Proterozoic than currently accepted, varying from <2 percent to over 15 percent (Large et al., 2019). The study also indicated that fluctuations in oxygen are greater and more frequent than other models during that time.

2.2 Pyrite Formation and Trace Element Incorporation

The formation of pyrite in sedimentary settings is discussed in detail by Rickard (2012), and a brief summary is presented here. Pyrite precipitation involves two important processes: nucleation of pyrite and mineral growth (Rickard, 2012). Direct precipitation of pyrite requires high levels of supersaturation, while growth on earlier pyrite, or other species, can occur at lower degrees of supersaturation (Harmandas et al., 1998; Rickard, 2012). This explains why pyrite is amenable to being overgrown by later pyrite generations, and each generation can preserve a trace element signature different from the earlier formed pyrite (Gregory et al., 2019b).

Pyrite forms via three main pathways. The first is reaction of H_2S with dissolved FeS, possibly via a solid $FeS_{(x)}$ precursor mineral (Rickard, 2012). The importance of precursor $FeS_{(x)}$ (especially mackinawite) has been called into question (Rickard and Morse, 2005), but it is the most commonly used starting material for pyrite formation experiments. An alternative to H_2S is the reaction of FeS species with a polysulfide complex to form pyrite (Rickard, 2012). Recent work has suggested that a tetrahedral FeS_{nano} solid precipitates prior to the traditionally invoked first solid $FeS_{(x)}$ phase, mackinawite (Matamoros-Veloza et al., 2019). This is particularly interesting because, over time, in the absence of oxygen, this phase loses electrons and some of the Fe^{2+} oxidizes to Fe^{3+} and forms a layered unsaturated structure. H_2S is then oxidized by the Fe^{3+} to form polysulfide complexes between the Fe layers (Matamoros-Veloza et al., 2019). This is important because this species is highly reactive and may be related to metal incorporation. Also, it provides a mechanism for Fe oxidation without external sources of oxidant; however, it should be noted that these phases require a low pH (≤ 4) (Matamoros-Veloza et al., 2019) which may not be present in many environments in which pyrite forms. Additionally, if the Fe(hydr)oxides are present, they can be sulfidized and replaced by pyrite (Peiffer et al., 2015). Currently the degree to which trace elements can be retained during the transfer from oxide to sulfide is unknown. However, seafloor ferromanganese nodules and crusts tend to be highly enriched in siderophile elements (thousands of ppm Co and Ni, hundreds of ppm Mo; Koschinsky and Hein, 2017) while As concentrations tend to be relatively low (tens of ppm). These ratios are atypical of most sedimentary pyrite so, if they are found, the results may deserve increased scrutiny before use in understanding past-ocean chemistry.

The rate of pyrite formation depends on several factors. The presence of cells, both dead and alive, has been found to increase the rates of mackinawite formation (Picard et al., 2018) and it is plausible that a similar effect would hold true with pyrite formation. The presence of certain trace elements may also affect the rate of pyrite formation. Morin et al. (2017) found that reaction rates of pyrite precipitation were significantly higher when Ni was present. However, this observation might be dependent on the reaction mechanism (Fe^{3+} reactant) rather than the presence of Ni because Swanner et al. (2019) observed no increase in the rate of pyrite precipitation with increases in Ni content when using mackinawite as a reactant.

Understanding the way in which trace elements are incorporated into pyrite is important if we are to utilize pyrite trace element content to understand past ocean chemistry. Unfortunately, the incorporation method of trace elements into pyrite is still not well understood. It has been shown that several trace elements can be adsorbed and incorporated into mackinawite (Morse and Arakaki, 1993), which can then convert into pyrite; however, pyrite does not necessarily form via a mackinawite precursor and thus transfer through a precursor phase may not be important in all settings. Furthermore, mackinawite has been shown to dissolve prior to pyrite precipitation so any bound trace elements would also be released into solution before pyrite formation (Rickard, 2012). The degree of trace metal pyritization (DTMP) of several transition metals, many of which have applications for understanding past ocean chemistry, appears to be related to the rate of their water exchange reactions and how those water exchange reactions rates compare to the rate of mackinawite formation (Morse and Luther, 1999). For example, Ni and Co have relatively slow water exchange kinetics and high DTMP, while Pb, Zn, and Cd have faster water exchange reactions and lower DTMP (Morse and Luther, 1999). Thus, Co and Ni are expected to be available to be incorporated into the structure of pyrite while the other elements are more likely to precipitate as different mineral phases, possibly as inclusions in pyrite. This is supported by Gregory et al. (2015a) and Gregory et al. (2014) who used time-resolved laser ablation output graphs to argue that Ni and Co tend to be incorporated into sedimentary pyrite by substitution for Fe in the pyrite structure, whereas Cd, Zn, and to a lesser extent Pb are commonly incorporated as micro-inclusions.

The most effort has been put into understanding the incorporation of As into pyrite, both because of its potential toxicity and its relation to gold in economic mineral deposits. While much of the work has been done on hydrothermal systems, some of the conclusions may have application to sedimentary pyrite as well. There are three ways that As can be incorporated into pyrite:

1 substitution of As (-I) for S(-II) (Reich and Becker, 2006)
2 substitution of As (III) for Fe(II) (Deditius et al., 2008; Qian et al., 2013)
3 amorphous As-Fe-S nano-inclusions (Deditius et al., 2008)

Which way As is incorporated is thought to affect the relative DTMP of other elements. In some cases, this may be quite important because As is among the most abundant trace elements in sedimentary pyrite (Gregory, et al., 2015a) and has one of the highest DTMP (Morse and Luther, 1999). When As(-I) substitutes for S(-II), it tends to increase the incorporation of elements with ionic radii similar to Fe(II), such as Co(II), Ni(II), Cu(II), and Zn(II) (Michel et al., 1994; Deditius et al., 2008). Conversely, when As(III) substitutes for Fe(II), charge balance requires that for every two As(III) ions there needs to be a vacancy left in the pyrite structure that allows for the incorporation of large ions such as Ag (I), Au(I), or Pb(II) (Cook and Chryssoulis, 1990; Fleet et al., 1997). Thus, when interpreting pyrite trace element data, it is important to be cognizant of the other elements that are enriched with the element of interest to ascertain whether an enrichment is solely due to a change in water chemistry or whether it may be related to increases in other elements, such as As.

Molybdenum is another element that has received significant scientific attention. This is largely because of its redox sensitive nature and long ocean residency time, which makes it very useful as a whole rock paleo-ocean chemistry proxy (Tribovillard et al., 2006). Molybdenum is often enriched in sulfidic sediments, however, it is not clear how it is incorporated into the sediments let alone into the pyrite phase. Chappaz et al., (2014) showed that while Mo is relatively enriched in pyrite it is also enriched within the host sediments and they hypothesized that it may be related to adsorption on organic matter. However, this has been disputed by Helz and Vorlicek (2019) who argued that organic ligands are not strong enough to bind the large amounts of Mo found in some shales. Rather they argue that Fe-Mo-S colloids are more likely pathways for Mo accumulation in sediments (Vorlicek et al., 2019). This would manifest itself in Mo being accumulated as nano- to micro-inclusions of this colloidal Mo-Fe-S particle or the minerals they transform into over time. Importantly this is contrary to predictions made by Gregory et al., (2015a) who argued, based on laser ablation time-resolved output graphs, that Mo was contained within the structure of sedimentary pyrite. There is some evidence – a lack of obvious Mo at the nano scale in the Mo-rich pyrite from the Cariaco Basin (Gregory et al., 2018) – that suggests that the predictions of Mo being incorporated into the pyrite structure are incorrect, which would tend to support the fixation of Mo in Fe-Mo-S colloids proposed by Vorlicek et al., (2018), which inter-grow with the forming pyrite as well as elsewhere in the sediments. This is important because it highlights the complexities of trace element incorporation in sedimentary pyrite, even in relatively well-studied systems.

The pyrite technique cannot be used on pyrite formed in sediments via replacement of Fe(hydr)oxides because this type of pyrite will inherit the trace element content from the original Fe(hydr)oxides (Dos Santos Afonso and Stumm, 1992). The precursor Fe(hydr)oxides formed in the sediments or water column may contain a similar set of trace element profiles as pyrite, but the partition coefficients will likely be vastly different, producing data unsuitable for paleo-ocean chemistry interpretations.

2.3 Using Pyrite Trace Element Chemistry to Check for Hydrothermal Overprints

One of the common arguments against interpretations using abundance of redox sensitive trace elements in whole rock is that the trace element enrichment is due to a hydrothermal source rather than past increases of the trace elements in the ocean. This can be hard to argue against because distal hydrothermal fluids can be cryptic and negatives are always hard to prove. Fortunately, there is much data generated by economic geologists who specialize in finding and characterizing hydrothermal pyrite. Through extensive pyrite data compilation and new pyrite studies, there is now sufficient hydrothermal and non-hydrothermal pyrite trace element data to allow for systematic pyrite classification. Sedimentary pyrite contains distinct trace element signatures and the criteria for sedimentary pyrite based on trace element ratios are listed in Table 1 (Gregory et al., 2015a). Using these criteria, it can be determined whether the unknown pyrite is from a sedimentary source (Mukherjee and Large, 2016;

Table 1 Trace element ratio
ranges indicative of pyrite formed
in sedimentary settings (Gregory
et al., 2015)

Ratio
0.01 < Co/Ni < 2
0.01 < Zn/Ni < 10
0.01 < Cu/Ni < 2
0.1 < As/Ni < 10
1 < Te/Au < 1000
As/Au > 200
Ag/Au > 2
Sb/Au > 100
Bi/Au > 1

Gregory et al., 2017). This technique has since been refined by coupling a database of known pyrite (five types of hydrothermal pyrite and sedimentary pyrite) trace element content with machine-learning algorithms. By doing this, sedimentary pyrite can be correctly identified 95 percent of the time per analysis and 100 percent of the time per sedimentary formation (Gregory et al., 2019a).

3 Materials and Methods

3.1 Pyrite Textures and Sample Preparation

Pyrite commonly forms in the pore waters of organic rich sediments or in sulfidic water. As both of these settings tend to be organic-rich, black shales are the main rock types in which the pyrite trace element proxy can be utilized. As with most geochemical proxies, the pyrite proxy gives only local conditions; to make sweeping generalizations about global ocean chemistry, samples from multiple locations representative of the same time interval should be obtained (Large et al., 2015).

The in-situ nature of the pyrite proxy technique simplifies the sample preparation procedures compared to many other proxies. No grinding and dissolution is required. The samples only need to be cut, mounted in epoxy, and polished with a final grit of 1 μm diamond paste. Extra care should be taken to avoid over-polishing, which can result in an uneven surface due to soft material being preferentially polished away. This is because uneven surfaces can result in inefficient transport of the ablated material into the ICPMS, which may adversely affect the LA-ICPMS analysis. Once polished mounts have been

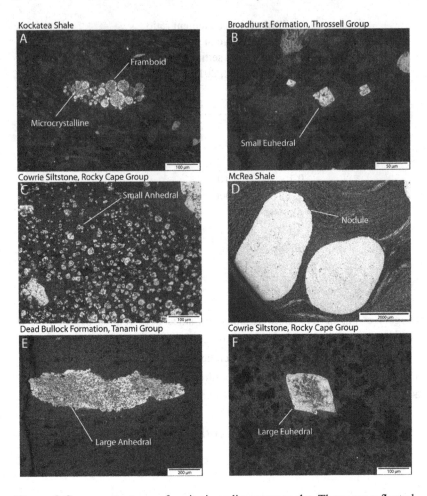

Figure 2 Common textures of pyrite in sedimentary rocks. These are reflected light images and samples have been etched with nitric acid. A) framboidal pyrite, B) small (< 15 μm) euhedral pyrite, C) small (< 15 μm) anhedral pyrite, D) nodular pyrite, E) large (> 15 μm) anhedral pyrite, F) large (> 15 μm) euhedral pyrite. Figure from Gregory et al., (2015a).

made, detailed petrographic observation should be undertaken to identify pyrites that are likely to have formed during the stage in the history of the sediments that is of interest. Nitric acid etching (Gregory et al., 2015a; Fig. 2, amongst many others) or NaOCl (Sykora et al., 2018) can be very useful in revealing the textures of the initial pyrite generations, as well as the relative timing of the formation of the different textured pyrite. Small (< 10 μm) pyrite framboids (Fig. 2A) are the most likely to have formed in the water column (Wilkin et al., 1996); thus, they are also the most likely to accurately reflect the trace element content of the ocean water they formed in and are the preferred textures to analyze. However, those framboids are difficult to analyze due to their small size, so clusters can be analyzed wherever available. Sometimes early, fine-grained pyrite framboids have later pyrite overgrowths (Wacey, et al., 2015; Gregory et al., 2019b); if such an early texture can be identified from later pyrite, they are also good targets for analysis. Late pyrite tends to be coarser grained and/or more euhedral, and are best avoided for past ocean chemistry studies.

If the goal is rather to track the changes through sediment diagenesis, analysis across later, rimming pyrite can give some useful information. This application again begins with detailed petrography including etching. Then either LA-ICPMS maps (Large et al., 2009; Genna and Gaboury, 2015, Gregory et al., 2015a; Steadman et al., 2015; Cooke et al., 2016) or a line of spot analyses (Gregory et al., 2019b) are conducted across the different identified textures to give an understanding of how pyrite chemistry (and by inference, pore fluid chemistry) varied during the formation of the pyrite. By coupling these analyses with in-situ S-isotope (SIMS or SHRIMP) analysis, the degree of openness of the system can be determined and an understanding of which diagenetic processes were occurring can be achieved.

3.2 Analytical Methods: LA-ICPMS

LA-ICPMS is now a common technique in many aspects of earth sciences, including dating of different mineral phases, understanding changes in chemistry of individual igneous and metamorphic minerals, determining fluid inclusion chemistry, understanding the chemistry of ore minerals, and now pore/ocean water chemistry via pyrite chemistry. Here, a summary of the method is presented and some potential issues and best practices for reporting data are highlighted. Interested readers are referred to Large et al. (2007), (2009), and Gregory et al. (2017), and references therein for details on the methodology of LA-ICPMS analysis of sedimentary pyrite.

The LA-ICPMS technique, essentially, utilizes laser energy to vaporize a small amount of the mineral of interest (in this case pyrite) in a He atmosphere. The resultant aerosol is then carried (an additional carrier gas, usually Ar, is usually added to help aerosol transport) to the mass spectrometer where it is ionized in a plasma and analyzed by the mass spectrometer. In addition to the target pyrite grains, approximately five spot analyses can also be analyzed on the black shale matrix, taking care to avoid small pyrite grains. This can later be used to correct for the matrix material that is often ablated with the target pyrite, especially with fine-grained pyrite. At the beginning and end of each set of analyses, a standard or multiple standards with known major and trace element composition is analyzed. The counts of the standard are compared to counts for an internal standard to convert counts of measured elements in the unknown pyrite to ppm (Fe is the most appropriate for pyrite; Longerich et al., 1996). Secondary standards, ideally pyrite or similar sulfide material, should also be interspersed with the unknown samples to check the accuracy of the analytical run. In addition to the elements of interest, major elements are also monitored to determine the quantity of matrix inclusions and appropriate standards to properly quantify them are utilized.

3.2.1 Data Processing

Once initial data reduction has been completed, the total abundance should be checked to ensure that the sum of all the elements is near 100 percent, using masses of the elements related to their most likely species. That is, you need to correct for elements that likely exist bound to oxygen as oxygen will not be quantified. If necessary, a correction factor should be applied to the resultant data. Once total abundances are calculated, the data are scrutinized for the amount of non-pyrite matrix minerals. It is recommended that if an analysis captured more than 20 percent matrix it is excluded; and that analyses that have less than 20 percent matrix are adjusted by using a correction factor that accounts for the trace elements held within the matrix inclusions. University of Tasmania researchers Leonid Danyushevsky and Sasha Stepanov specialize in LA-ICPMS data reduction algorithms and they suggest segmenting the area of integration during data processing into five groups. The chalcophile and siderophile elements are then plotted against S and linear regression equations are calculated. The final concentrations are then calculated with the S content that brings the total siderophile and chalcophile concentration up to 100 percent (see Gregory et al., 2017; Stepanov et al., 2020, for details).

In addition to understanding the changing trace element concentrations captured by sedimentary pyrite, the data obtained through LA-ICMPS can also shed light on how trace elements are held in pyrite (i.e., as micro-inclusions or bound within the pyrite structure). This is done by close inspection of time resolved laser ablation output graphs (Fig. 3). When there are irregular sharp peaks or drops in the total counts, it is likely that the laser beam is ablating through a single or multiple inclusions. An example of this is the two Au peaks in Fig. 3D and Pb and Bi peaks in Fig. 3E and 3 F respectively. All of these peaks are likely due to the laser beam ablating through micro-inclusions of minerals that contain the respective elevated elements in them. In contrast, Ni, Co, and As tend to parallel the counts of Fe in each of the examples in Fig. 3, indicating

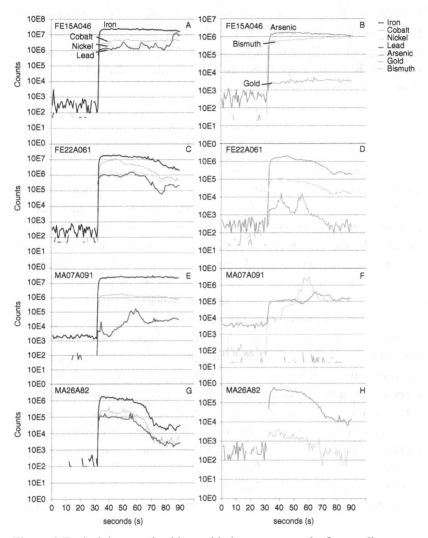

Figure 3 Typical time resolved laser ablation output graphs from sedimentary pyrite from 4 different samples. Most trace elements are relatively even counts throughout ablation, indicating consistent trace element content. A few analyses have peaks in Pb and Bi (e and f) indicating an inclusion of a mineral containing both these elements and two peaks in Au are evident in d, indicating inclusions of Au. Figure from Gregory et al., (2015a).

that these minerals are held within the pyrite structure. However, LA-ICPMS is not considered sensitive enough to determine whether a given element is held with the structure of pyrite or within evenly distributed nano-inclusions (Gregory et al., 2015a). As outlined, LA-ICPMS analyses of the sedimentary matrix are required in order to accurately resolve the trace element content in pyrite. This means that elements commonly present in the matrix, and not in the pyrite (e.g., Sn, W, Ti, Zr, U, Th, Cr, and the REE), can be determined for each sample and be used to support other useful proxies such as Th/U, Th/Cr, Eu/Eu* (Large et al., 2018).

3.2.2 Reporting Standards

It is important to report the standards used, the concentration of all elements in the standards, and the known concentrations in any secondary standards Barnes (2019). This will allow correction of the data if, at a later date, new analytical results show different concentrations of some elements of the standards. In addition to reporting the data required to reassess the data generated at a later date, it is also important to report a wide suite of different elements to facilitate data compilations: we recommend reporting a minimum of As, Ni, Pb, Cu, Co, Mn, Sb, Zn, Se, Mo, Ag, Bi, Te, Cd, Au, Sn, and W (Gregory et al., 2015a; 2019a), even if the values are below detection limits.

The amount of analytical error for each element will vary depending on the analytical conditions for a given spot. Changing background counts, size of laser spot, counting time on a given element, and time interval used in data reduction will all affect the error of a given analysis. However, having a general idea of what percent error can be expected can aid in planning analytical procedures and what time interval of measurements should be taken on a given element to obtain a desired percent error for the given element. In Table 2, the median error for the elements commonly of interest in sedimentary pyrite analyses from the Doushantuo Formation, China, are given (Gregory et al., 2017). These are of small pyrite that are ideal for past ocean chemistry studies. Generally, errors are between 10 and 20 percent, with

Table 2 Median % errors for elements commonly held in pyrite from samples analyzed in Gregory et al., 2017

Element	Mn	Co	Ni	Cu	Zn	As	Se	Mo	Ag	Sb	Te	Au	Tl	Pb	Bi
Median % error	27	8.9	6.5	12	60	6.6	7.2	18	12	9.5	15	19	17	9.1	14
Associated concentration* (ppm)	118	50	300	522	39	250	61	9.3	3.9	40	2.8	0.08	2.5	116	2.3

* By associated concentration we mean the concentration of the given element for the sample with the median error estimate. This is done because error estimates are related to the total counts for the element of interest.

the higher errors usually belonging to elements with lower abundance. For example, Au has a median % error of 19 percent but the sample with this error had a measured value of 0.08 ppm Au. Conversely, Ni has a low median % error (6.5) but with a relatively high concentration (300 ppm). Some elements do deviate from this trend, such as Zn (median % error 60) and Mn (median % 27). This may be due to these elements often being contained within micro-inclusions (Gregory et al., 2015a) in pyrite resulting in variable concentrations being integrated. Thus particular care must be taken when interpreting results from these elements.

3.2.3 Potential Issues with LA-ICPMS Analyses

One of the benefits of LA-ICPMS analyses is that it does not require dissolution of the sample and thus difficulties in dissolving geologic samples and the dilution that accompanies any dissolution can be avoided. However, this also means that no separation chemistry can be conducted prior to analysis in the ICPMS, thus possibility of mass peak overlaps can be high and elements that could potentially cause peak overlaps on elements of interest should be monitored. For example, $^{103}Rh^+$ has a mass charge ratio of +103, but so do $^{206}Pb^{2+}$, $^{87}Rb^{16}O^+$, $^{87}Sr^{16}O^+$, $^{87}Rb^{16}O^+H^+$, and $^{63}Cu^{40}Ar^+$ (Danyushevsky, 2019) – so care must be taken when interpreting the results of LA-ICPMS analyses. Because every sample will have different abundances of the different elements, it is difficult to predict potential interferences and analysts are encouraged to investigate pertinent ICP-MS literature, such as May and Wiedmeyer (1998), and also critically examine the potential interferences in their unique samples prior to analysis. For paleo-ocean chemistry studies, As, Mo, Co, Ni, and Se can be among the most important and each has its own potential interferences. Selenium can be particularly problematic because ^{74}Se, ^{76}Se, ^{78}Se, and ^{80}Se, which include the most abundant Se isotopes (^{78}Se abundance of 23.53 percent and ^{80}Se abundance of 49.82 percent) and all have interferences with different polyatomic Ar species. For example, ^{74}Se is interfered by $^{36}Ar^{38}Ar^+$, ^{76}Se is interfered by $^{40}Ar^{36}Ar^+$ and $^{38}Ar^{38}Ar^+$, ^{78}Se is interfered by $^{40}Ar^{38}Ar^+$, and ^{80}Se is interfered by $^{40}Ar^{40}Ar^+$ (Tan and Horlick, 1986; Longbottom et al., 1994; May and Weider, 1998). These can be particularly problematic because Ar is a carrier gas in most LA-ICPMS experimental apparatus and thus production of these species can be difficult to avoid. Nickel also has potential interferences on Ar containing polyatomics such as $^{40}Ar^{18}O^+$ and $^{40}Ar^{17}O^1 H^+$ on ^{58}Ni. For Ni there may also be interferences from polyatomic molecules that may come from matrix material in the rock samples such as $^{23}Na^{35}Cl^+$, $^{40}Ca^{18}O^+$, $^{40}Ca^{18}O^1 H^+$, and $^{42}Ca^{16}O^+$ for ^{58}Ni and $^{23}Na^{37}Cl^+$, $^{44}Ca^{16}O^+$, and $^{43}Ca^{16}O^1 H^+$ for ^{60}Ni

(McLaren et al., 1985; Tan and Horlick, 1986; Plantz et al., 1989; Evans and Giglio, 1993; Reed et al., 1994; May and Weidmeyer, 1998). These could be particularly important interferences in modern ocean sediments for NaCl or carbonate rich sediments for the CaO interferences. Molybdenum has a number potential interference for each of its isotopes that could be derived by phyllosilicate minerals in the matrix material. These include $^{39}K_2^{16}O^+$, for ^{94}Mo; $^{40}Ar^{39}K^{16}O^+$ for ^{95}Mo; $^{39}K^{41}K^{16}O^+$ for ^{96}Mo; $^{40}Ar^{41}K^{16}O^+$ and $^{40}Ca_2^{16}O^1 H^+$ for ^{97}Mo; and $^{41}K_2O^+$ (Beary and Paulson, 1993; Vandecasteele et al., 1993, May and Weidmeyer, 1998). One possible strategy for determining whether there is a significant effect on the abundance of one element is by monitoring its other isotopes because, while they may all be affected as shown, the isotopes should not be equally affected so if each isotope gives the same value for its elements abundance it can be accepted as being reasonably accurate. This will not work for elements such as Co and As as they only have one isotope. Like Ni, Co and As can have interferences from carbonate rich matrixes because there are interferences from $^{43}Ca^{18}O^+$ and $^{42}Ca^{18}O^1 H^+$ on ^{59}Co and $^{43}Ca^{16}O_2^+$ on ^{75}As (Tan and Horlick, 1986; Evans and Giglio, 1993; Campbell et al., 1994; Longbottom et al., 1994). The risk of incorrectly identified elements can be limited by monitoring the potentially interfering elements and checking whether increases in one are coupled with increases in the other, which could show an interference rather than a real elemental enrichment.

Due to their fine-grained nature, black shales can have abundant volatile molecules adsorbed to them and the volatiles can desorb during LA-ICPMS analysis. This can lead to high background counts and some unanticipated peak overlaps. To limit this effect, samples and standards are recommended to be kept under vacuum prior to analysis (Danyushevsky, 2019). If time between sample preparation and analysis is limited, then time in a low temperature vacuum oven can speed up this process.

3.2.4 Summarizing Data

Pyrite trace element content has variable distributions. It is almost never normally distributed (Gregory et al., 2019a), and, as such, should not be summarized using arithmetic means and standard deviations. More commonly, trace element content approximates a log normal distribution. If this is the case, then the geometric mean and multiplicative standard deviation can be used effectively. However, it is important to understand the statistical distribution of a dataset. Check whether there is indeed a log normal distribution, as this is not always the case, especially with low abundance elements, such as gold. If the data is not log-normally distributed, then the median value and the median absolute deviations can be used instead (Gregory et al., 2019a). We stress that the use of the correct statistical quantities is very important when reporting LA-ICPMS analyses of pyrite because it is not uncommon to have a multiple order of magnitude spread in the data. This means that single far outliers can significantly skew the mean and the calculated arithmetic standard deviation will represent far less than 68 percent of the data and/or the interval defined by the standard deviation will go into the negative, a physical impossibility.

4 Case Study

4.1 Comparison between Pyrite Trace Element Chemistry and Traditional Paleo-Ocean Proxies

To test the pyrite trace element proxy at the section scale, Gregory, et al. (2017) selected the Wuhe section of the Doushantuo Formation, China. This section was chosen because it was deposited under persistently euxinic conditions as determined by a combination of Fe-speciation analyses (Sahoo et al., 2016) and pyrite framboid size analysis (Wang et al., 2012). The well-established depositional setting eliminates the potential complexity of changing water column redox conditions. Additionally, the section had already been analyzed by Sahoo et al. (2016) for whole rock redox sensitive trace element abundance (i.e., Mo, V, U, and Re) and these data had been interpreted to show three distinct short-lived oxygenation events prior to the NOE (see Tribovillard et al., 2006, for a description of these whole rock proxies). Sahoo et al., (2016) also presented $\delta^{34}S$ values for the section, which informs on how the sulfate budget changed during deposition of the Wuhe section.

Gregory et al. (2017) found that most of the chalcophile elements are elevated in pyrite at the same horizon as the redox sensitive elements are elevated in whole rock. Importantly, the pyrite proxy also appeared to be more sensitive than the whole rock proxy, with Mo enrichment factors of 6.3 to 23.9 in intervals interpreted to have experienced oxygenation events (Fig. 4). This provides support for the conclusions of Sahoo et al., (2016) and also provides support for the use of the pyrite trace element proxy for sedimentary sections deposited under euxinic conditions. A larger number of chalcophile elements than expected were found to correctly predict periods of increased oxygenation. This is interpreted to be because of the chalcophile elements' affinity for bonding with sulfur. During times of large amounts of euxinia in the basin, pyrite is formed over a wide area resulting in widespread drawdown of chalcophile elements and a general depletion in these elements in the basin as a whole (Gregory et al., 2017). The relatively high $\delta^{34}S$ supports this interpretation as at times of widespread pyrite formation sulfur becomes depleted and residual sulfate becomes more enriched in the heavy

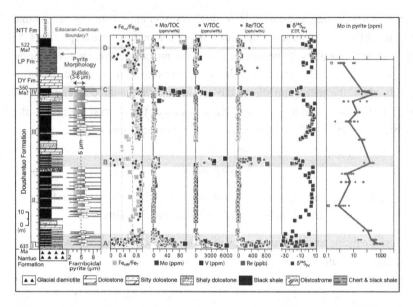

Figure 4 Comparison of whole rock geochemistry (Sahoo et al., 2016) and pyrite trace element content at the section scale. Note the increase of Mo and most of the chalcophile elements during the same intervals that there are increases of redox sensitive trace elements. This demonstrates how Mo and chalcophile trace element composition of pyrite can be used in a similar fashion to bulk analyses of redox sensitive trace elements. Figure from Gregory et al., (2017).

[34]S because as the lighter [32]S is preferentially used up. Then at times when oxygen increases the area of the seafloor under an euxinic water column decreases which limits production of pyrite and slows the drawdown of both [32]S and chalcophile trace elements (Gregory et al., 2017, Sahoo et al., 2016). Thus where the water column is still euxinic, chalcophile trace elements in pyrite increase and δ^{34}S decreases. Therefore, not only does the pyrite proxy reflect traditional proxies, it allows far more trace elements to be utilized in paleo-ocean chemistry reconstructions (Gregory et al., 2017).

5 Future Prospects

While promising, there are still many unresolved issues with the pyrite trace element proxy. Most of the studies thus far have focused on producing compilations of pyrite data through geologic time (Large et al., 2014; Gregory et al., 2015a; Large et al., 2015). While these compilations tend to match well with existing whole rock chemistry compilations (Scott et al., 2008; Konhauser et al.,

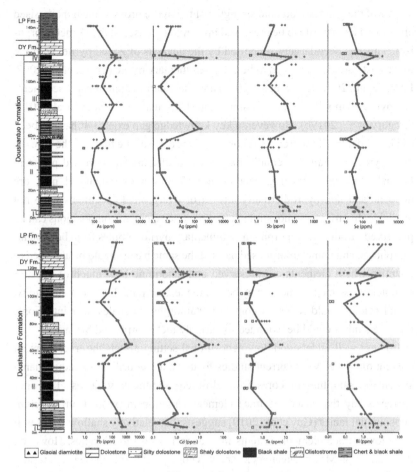

Figure 4 (cont.)

2009; Partin et al., 2013), there are still many questions that are unresolved. First, pyrite can form in many environments, within sediments under oxic or anoxic water columns or in the water column in euxinic settings. To what degree the trace element content of pyrite varies in each of these environments and how pyrite chemistry should be compared when sourced from different environments remains untested. Detailed analyses of pyrite forming in modern settings still need to be conducted to fully understand the potential variability of pyrite formed under these different conditions and to produce a work flow to refine the proxy. Furthermore, these studies should be conducted through ocean drilling program cores to ascertain how the trace element content of pyrite changes as different sulfate reduction mechanisms become more prevalent (i.e., thermo-chemical sulfate reduction).

One of the greatest potential strengths of the pyrite proxy is the hypothesised ability of TEs in pyrite to be preserved up to mid greenschist facies metamorphism. This has been largely supported by the analysis of zoned pyrite, with the innermost core being argued to be early sedimentary pyrite (Large, et al., 2007, Large, et al., 2009, Large, et al., 2011) and the identification of early sedimentary pyrite framboidal textures using nanoSIMS analysis (Wacey et al., 2015; Gregory et al., 2019b). However, a key knowledge gap exists – it has not been established that the trace element content of the early pyrite found in the cores of later pyrite generations actually matches that of unmetamorphosed pyrite in basinal sediments. Existing data is limited to studies conducted near ore deposits and non-hydrothermally altered or overprinted specimens were not available. In order to demonstrate that pyrite trace element compositions can be preserved through metamorphism, sedimentary pyrite samples from both metamorphosed and unmetamorphosed parts of the same basin should be analyzed to quantify which elements are preserved and determine how much of the not-completely retained elements may be lost to metamorphic fluids. For simplicity, initial studies could focus on contact metamorphosed shales near intrusions, then the studies could be extended to variably metamorphosed basins.

It is not well understood how depositional setting affects the trace element content of pyrite. Most current studies focus on a few drill cores rather than a systematic sampling of cores from different sedimentary facies. The only current study that discusses trace element variation in pyrite from different depositional facies (Guy et al., 2010) suggests that relatively shallow facies will have lower trace element contents than deeper depositional facies. However, this study is on Mesoarchean rocks and it is not clear whether these results apply to post-Archean time periods when atmospheric oxygen levels are much higher and thus depositional flux are significantly different. Also, at these times, pyrite deposition might also be more related to volcanic processes than more recent times (i.e., post-GOE) when sulfate levels are much higher (Olson et al., 2019). To test how pyrite trace element content varies across a basin, a systematic set of samples can be taken from contemporaneous, sedimentary samples from different facies from a well-understood, non-metamorphosed basin. These data could then be used to assess the potential variability of the pyrite trace element content due to depositional facies. Initial steps have been made on this (Mukherjee et al., 2018b) with Cambrian shales in Australia but the results are yet to be published in full and the procedure should be repeated on multiple basins at different periods in Earth History.

Finally, little experimental work has been done to assess partition coefficients of different trace elements into pyrite at low temperatures. Swanner et al., 2019 investigated Co and Ni incorporation into pyrite and which is more representative of water chemistry from which it formed, but few other similar studies have been published. Thus, to fully understand the relationship between pyrite chemistry and its relationship to the chemistry of the water from which it formed, pyrite should be produced in solutions with different trace element compositions and the trace element content of the resultant pyrite should be analysed to obtain a set of partition coefficients.

While there is still more work required to refine the method, using pyrite trace element chemistry has many potential advantages. The ability to be preserved through metamorphism is perhaps the most important, especially in the Precambrian where researchers are frustrated by a vanishingly small number of basins that can be utilized to understand ocean and atmospheric chemistry at important periods of the evolution of our planet. The pyrite technique may provide glimpses into basins that currently are too high metamorphic grade for traditional proxies to be applied. The number of elements that can be accurately analyzed by LA-ICPMS and appear to reflect past ocean chemistry can give support to conclusions because they allow for the application of several lines of evidence to support interpretations. Finally, the ability of pyrite chemistry to identify the presence or lack of an overprint by hydrothermal fluids can help to resolve several disagreements whether a given redox sensitive trace element enrichment is due to changes in past ocean chemistry or later hydrothermal and/ or metamorphic overprint.

Key References

This volume provides an extensive overview of how pyrite forms in sedimentary settings.

Rickard, D. (2012) *Sulfidic Sediments and Sedimentary Rocks* (Elsevier) p. 801.

These papers are among the first to investigate sedimentary pyrite trace element abundance using methods other than LA-ICPMS.

Berner, Z. A., Puchelt, H., Nöltner, T. and Kramar, U. T. Z. (2013) Pyrite geochemistry in the Toarcian Posidonia Shale of south-west Germany: Evidence for contrasting trace-element patterns of diagenetic and syngenetic pyrites. *Sedimentology*, 60 548–573.

Huerta-Diaz, M. A., and Morse, J. W. (1990) A quantitative method for determination of trace metal concentrations in sedimentary pyrite. *Marine Chemistry* 29, 119–144.

Huerta-Diaz, M. A., and Morse, J. W. (1992) Pyritization of trace metals in anoxic marine sediments. *Geochimica et Cosmochimica Acta* 56, 2681–2702.

These papers were among the first to compile LA-ICPMS trace element data of pyrite and showed that it matches existing whole rock studies.

Gregory, D. D., Large, R. R., Halpin, J.A., Baturina, E. L., Lyons, T.W., Wu, S., Danyushevsky, L., Sack, P. J., Chappaz, A., and Maslennikov, V. V. (2015a) Trace Element Content of Sedimentary Pyrite in Black Shales. *Economic Geology* 110, 1389–1410.

Large, R. R., Halpin, J. A., Danyushevsky, L. V., Maslennikov, V. V., Bull, S. W., Long, J. A., Gregory, D. D., Lounejeva, E., Lyons, T. W., and Sack, P. J. (2014) Trace element content of sedimentary pyrite as a new proxy for deep-time ocean–atmosphere evolution. *Earth and Planetary Science Letters* 389, 209–220.

These papers provide examples where the proxy was used to identify oxygenation events at different times in Earth History.

Gregory, D. D., Large R. R., Halpin, J. A., Steadman, J. A., Hickman, A. H., Ireland, T. R., and Holden, P. (2015b) The chemical conditions of the late Archean Hamersley basin inferred from whole rock and pyrite geochemistry with Δ 33 S and δ 34 S isotope analyses. *Geochimica et Cosmochimica Acta* 149, 223–250.

Gregory, D. D., Lyons, T. W., Large, R. R., Jiang, G., Stepanov, A. S., Diamond, C. W., Figueroa, M. C., and Olin, P. (2017) Whole rock and discrete pyrite geochemistry as complementary tracers of ancient ocean chemistry: An example from the Neoproterozoic Doushantuo Formation, China. *Geochimica et Cosmochimica Acta* 216, 201–220.

Mukherjee, I., and Large, R. R. (2016) Pyrite trace element chemistry of the Velkerri Formation, Roper Group, McArthur Basin: Evidence for atmospheric oxygenation during the Boring Billion. *Precambrian Research* 281, 13–26.

This paper shows how trace element content of pyrite can be used to distinguish between sedimentary pyrite and hydrothermal pyrite.

Gregory, D. D., Large, R. R., Cracknell, M. J., Kuhn, S., Maslennikov, V. V., Belousoc, I. A., McGoldrich, P., Fabris, A., Baker, M. J., Fox, N., and Lyons, T. W. (2019a) Prediction of ore deposit style from Random Forest analysis of LA-ICPMS analyses of pyrite. *Economic Geology.*

This paper shows how the trace element content of pyrite in pyrite nodules can be used to obtain information of pore water chemistry during diagenesis.

Gregory D. D., Mukherjee, I., Large, R. R., Lyons, T. W., Stepanov, A., Avila, J., Olson, S. L., Ireland, T. R., Olin, P. H., and Danyushevsky, L. V. (2019b) The formation mechanism of sedimentary pyrite nodules determined by trace element and sulfur isotope microanalysis. *Geochimica et Cosmochimica Acta* 259, 53–68.

This paper uses trace element ratios of through geologic time and currently accepted oxygen levels to model atmospheric oxygen content.

Large, R. R., Mukherjee, I., Gregory, D., Steadman, J., Corkrey, R., and Danyushevsky, L. V. (2019). Atmosphere oxygen cycling through the Proterozoic and Phanerozoic. *Mineralium Deposita* 126, 1–22.

Additional References

Baldwin, B., and Butler, C. O. (1985) Compaction curves, *AAPG bulletin*, 69, 622–626.

Barnes, S. -J., (2019) *Sulfide minerals in igneous systems: Laser ablation applied to ore deposits*, GAC MAC short course notes, Quebec City, p. 66.

Beary, E., and Paulsen, P. (1993) Selective application of chemical separations to isotope dilution inductively coupled plasma mass spectrometric analyses of standard reference materials. *Analytical Chemistry* 65, 1602–1608.

Belousov, I., Large, R. R., Meffre S., Danyushevsky L. V., Steadman, J., and Beardsmore, T. (2016) Pyrite compositions from VHMS and orogenic Au deposits in the Yilgarn Craton, Western Australia: Implications for gold and copper exploration. *Ore Geology Reviews* 79, 474–499.

Berner, Z., Pujol, F., Neumann, T., Kramar, U., Stüben, D., Racki, G., and Simon, R. (2006) Contrasting trace element composition of diagenetic and syngenetic pyrites: implications for the depositional environment. *Geophysical Research Abstracts.*

Blamey, N. J., Brand, U., Parnell, J., Spear, N., Lécuyer, C., Benison, K., Meng, F., and Ni, P. (2016) Paradigm shift in determining Neoproterozoic atmospheric oxygen. *Geology* 44, 651–654.

Campbell, M. J., Demesmay, C., and Ollé, M. (1994) Determination of total arsenic concentrations in biological matrices by inductively coupled plasma mass spectrometry. *Journal of Analytical Atomic Spectrometry* 9, 1379–1384.

Chappaz, A., Lyons, T. W., Gregory, D. D., Reinhard, C. T. Gill, B. C., Li, C., and Large, R. R. (2014) Does pyrite act as an important host for molybdenum in modern and ancient euxinic sediments? *Geochimica et Cosmochimica Acta* 126, 112–122.

Cook, N. J., and Chryssoulis, S.L. (1990) Concentrations of invisible gold in the common sulfides. *The Canadian Mineralogist* 28, 1–16.

Cook, N., Ciobanu, C., George, L., Zhu, Z. Y., Wade, B., and Ehrig, K. (2016) Trace element analysis of minerals in magmatic-hydrothermal ores by laser ablation inductively-coupled plasma mass spectrometry: Approaches and opportunities. *Minerals* 6, 111.

Danyushevsky, L. (2019) *Overview of LA-ICPMS: Laser ablation applied to ore deposits*, GAC MAC short course notes, Quebec City, p.66.

Deditius, A. P., Utsunomiya, S., Renock, D., Ewing, R. C., Ramana, C. V., Becker, U., and Kesler, S. E. (2008) A proposed new type of arsenian pyrite: Composition, nanostructure and geological significance. *Geochimica et Cosmochimica Acta* 72, 2919–2933.

Diamond, C. W., and Lyons, T. W. (2018) Mid-Proterozoic redox evolution and the possibility of transient oxygenation events. *Emerging Topics in Life Sciences* 2, 235–245.

Diamond, C. W., Planavsky, N. J., Wang, C., and Lyons, T. W. (2018) What the~ 1.4 Ga Xiamaling Formation can and cannot tell us about the mid-Proterozoic ocean. *Geobiology* 16, 219–236.

Dos Santos Afonso, M., and Stumm, W. (1992) Reductive dissolution of iron (III) (hydr)oxides by hydrogen sulfide. *Langmuir* 8, 1671–1675.

Evans, E. H., and Giglio, J. J. (1993) Interferences in inductively coupled plasma mass spectrometry. A review. *Journal of Analytical Atomic Spectrometry* 8, 1–18.

Fleet, M. E., and Mumin, H. A. (1997) Gold-bearing arsenian pyrite and marcasite and arsenopyrite from Carlin Trend gold deposits and laboratory synthesis. *American Mineralogist* 82, 182–193.

Gadd, M. G., Layton-Matthews, D., Peter, J. M., and Paradis, S. J. (2016) The world-class Howard's Pass SEDEX Zn-Pb district, Selwyn Basin, Yukon.

Part I: trace element compositions of pyrite record input of hydrothermal, diagenetic, and metamorphic fluids to mineralization. *Mineralium Deposita* 51, 319–342.

Gadd, M. G., Peter, J. M., Jackson, S. E., Yang, Z., and Petts, D. (2019) Platinum, Pd, Mo, Au and Re deportment in hyper-enriched black shale Ni-Zn-Mo-PGE mineralization, Peel River, Yukon, Canada. *Ore Geology Reviews* 107, 600–614.

Genna, D., and Gaboury, D. (2015) Deciphering the hydrothermal evolution of a vms system by LA-ICP-MS using trace elements in pyrite: An example from the Bracemac-McLeod Deposits, Abitibi, Canada, and implications for exploration. *Economic Geology* 110, 2087–2108.

Guy, B., Beukes N., and Gutzmer, J. (2010) Paleoenvironmental controls on the texture and chemical composition of pyrite from non-conglomeratic sedimentary rocks of the Mesoarchean Witwatersrand Supergroup, South Africa. *South African Journal of Geology* 113, 195–228.

Gregory D.D., Chappaz A., Atienza, N., Taylor, S., Perea D., Kovarik L., and Lyons T.W. Is pyrite an important sink for Mo? Evidence from XAFS, TEM and APT analyses of pyrite. Goldschmidt, Honolulu, Hawaii, United States, June, 21–26.

Gregory, D., Meffre, S., and Large, R. (2014) Comparison of metal enrichment in pyrite framboids from a metal-enriched and metal-poor estuary. *American Mineralogist* 99, 633–644.

Gregory, D., Perea, D., Taylor, S., Kovarik, L., Owens, J., and Lyons, T. (2019) The formation of pyrite framboids: A view from TEM and APT. Goldschmidt, Barcelona, Spain, August, 18–23.

Harmandas, N. G., Navarro Fernandez, E., and Koutsoukos, P. G. (1998) Crystal growth of pyrite in aqueous solutions. Inhibition by organophosphorus compounds. *Langmuir* 14, 1250–1255.

Harrold, T. W. D., Swarbrick, R. E., and Goulty, N. R. (1999) Pore pressure estimation from mudrock porosities in Tertiary basins, southeast Asia. *AAPG bulletin* 83, 1057–1067.

Helz, G. R., and Vorlicek, T. P. (2019) Precipitation of molybdenum from euxinic waters and the role of organic matter. *Chemical Geology* 509, 178–193.

Keith, M., Smith, D. J., Jenkin, G .R. T., Holwell, D. A., and Dye, M. D. (2018) A review of Te and Se systematics in hydrothermal pyrite from precious metal deposits: Insights into ore-forming processes. *Ore Geology Reviews* 96, 269–282.

Koschinsky, A. and Hein, J.R. (2017). Marine ferromanganese encrustations: Archives of changing oceans. Elements 13, 177–182.

Konhauser, K. O., Pecoits, E., Lalonde, S. V., Papineau, D., Nisbet, E. G., Barley, M. E., Arndt, N. T., Zahnle, K., and Kamber, B. S. (2009) Oceanic nickel depletion and a methanogen famine before the Great Oxidation Event. *Nature* 458, 750–753.

Large, R. R., Maslennikov, V. V., Robert, F., Danyushevsky, L. V., and Chang, Z.S. (2007) Multistage sedimentary and metamorphic origin of pyrite and gold in the giant Sukhoi Log deposit, Lena gold province, Russia. *Economic Geology* 102, 1233–1267.

Large, R. R., Danyushevsky, L., Hollit, C., Maslennikov, V., Meffre, S., Gilbert, S., Bull, S., Scott, R., Emsbo, P., Thomas, H., Singh, B., and Foster, J. (2009) Gold and trace element zonation in pyrite using a laser imaging technique: Implications for the timing of gold in orogenic and Carlin-style sediment-hosted deposits. *Economic Geology* 104, 635–668.

Large, R. R., Bull, S. W., and Maslennikov, V. V. (2011) A carbonaceous sedimentary source-rock model for Carlin-type and orogenic gold deposits. *Economic Geology* 106, 331–358.

Large, R. R., Halpin, J. A., Lounejeva, E., Danyushevsky, L. V., Maslennikov, V. V., Gregory, D., Sack, P. J., Haines, P. W., Long, J. A., and Makoundi, C. (2015) Cycles of nutrient trace elements in the Phanerozoic ocean. *Gondwana Research* 28, 1282–1293.

Large, R. R., Mukherjee, I. Zhukova, I., Corkrey, R., Stepanov, A., and Danyushevsky, L .V. (2018) Role of upper-most crustal composition in the evolution of the Precambrian ocean–atmosphere system. *Earth and Planetary Science Letters* 487, 44–53.

Longbottom, J., Martin, T., Edgell, K., Long, S., Plantz, M., Warden, B., Baraona, R., Bencivengo, D., Cardenas, D.,and Faires, L. (1994) Determination of trace-elements in water by inductively-coupled plasma-mass spectrometry-collaborative study. *Journal of AOAC International* 77, 1004–1023.

Longerich, H. P., Jackson, S. E., and Günther, D. (1996) Inter-laboratory note: Laser ablation inductively coupled plasma mass spectrometric transient signal data acquisition and analyte concentration calculation. *Journal of Analytical Atomic Spectrometry* 11, 899–904.

Lyons, T. W., Werne, J. P., Hollander, D. J., and Murray, R. W. (2003) Contrasting sulfur geochemistry and Fe/Al and Mo/Al ratios across the last oxic-to-anoxic transition in the Cariaco Basin, Venezuela. *Chemical Geology* 195, 131–157.

Matamoros-Veloza, A., Cespedes, O., Johnson, B.R., Stawski, T.M., Terranova, U., de Leeuw, N.H. and Benning, L.G. (2018). A highly reactive precursor in the iron sulfide system. Nature communications 9, 3125.

May, T. W., and Wiedmeyer, R. H. (1998) A table of polyatomic interferences in ICP-MS. *Atomic Spectrometry* 19, 150–155.

McLaren, J., Mykytiuk, A., Willie, S., and Berman, S. (1985) Determination of trace metals in seawater by inductively coupled plasma mass spectrometry with preconcentration on silica-immobilized 8-hydroxyquinoline. Analytical Chemistry 57, 2907–2911.

Michel, D., Giuliani, G., Olivo, G. R., and Marini, O. J. (1994) As growth banding and the presence of Au in pyrites from the Santa Rita gold vein deposit hosted in Proterozoic metasediments, Goias State, Brazil. *Economic Geology* 89, 193–200.

Morin, G., Noël, V., Menguy, N., Brest, J., Baptiste, B., Tharaud, M., Ona-Nguema, G., Ikogou, M., Viollier, E., and Juillot, F. (2017) Nickel accelerates pyrite nucleation at ambient temperature. *Geochemical Perspectives Letters* 5, 6–11.

Morse, J. W., and Arakaki, T. (1993) Adsorption and coprecipitation of divalent metals with mackinawite (FeS). *Geochimica et Cosmochimica Acta* 57, 3635–3640.

Morse, J. W., and Luther, G. W. (1999) Chemical influences on trace metal-sulfide interactions in anoxic sediments. *Geochimica et Cosmochimica Acta* 63, 3373–3378.

Mukherjee, I., and Large, R. (2017) Application of pyrite trace element chemistry to exploration for SEDEX style Zn-Pb deposits: McArthur Basin, Northern Territory, Australia. *Ore Geology Reviews* 81, 1249–1270.

Mukherjee, I., Large, R. R., Corkrey, R., and Danyushevsky, L. V. (2018a) The Boring Billion, a slingshot for complex life on Earth. *Scientific Reports* 8, 4432.

Mukherjee I., Large, R., Corkrey, R., Willink, R., and Stepanov, A. (2018b) How robust is sedimentary pyrite trace element geochemistry as a geochemical proxy? *Goldschmidt Abstracts* 2018, 1829.

Olson, S. L., Ostrander, C. M., Gregory, D. D., Roy, M., Anbar, A. D., and Lyons, T. W. (2019) Volcanically modulated pyrite burial and ocean–atmosphere oxidation. *Earth and Planetary Science Letters* 506, 417–427.

Partin, C. A., Bekker, A., Planavsky, N. J., Scott, C. T., Gill, B. C., Li, C., Podkovyrov, V., Maslov, A., Konhauser, K. O., and Lalonde, S. V. (2013) Large-scale fluctuations in Precambrian atmospheric and oceanic oxygen levels from the record of U in shales. *Earth and Planetary Science Letters* 369, 284–293.

Peiffer, S., Behrends, T., Hellige, K., Larese-Casanova, P., Wan, M., and Pollok, K. (2015) Pyrite formation and mineral transformation pathways upon sulfidation of ferric hydroxides depend on mineral type and sulfide concentration. *Chemical Geology* 400, 44–55.

Picard, A., Gartman, A., Clarke, D. R., and Girguis, P. R. (2018) Sulfate-reducing bacteria influence the nucleation and growth of mackinawite and greigite. *Geochimica et Cosmochimica Acta* 220, 367–384.

Piper, D. Z., and Dean, W. E. (2002) Trace-element deposition in the Cariaco Basin, Venezuela Shelf, under sulfate-reducing conditions: A history of the local hydrography and global climate, 20 ka to the present. US Geological Survey.

Pisarzowska, A., Berner, Z. A., and Racki, G. (2014) Geochemistry of Early Frasnian (Late Devonian) pyrite-ammonoid level in the Kostomłoty Basin, Poland, and a new proxy parameter for assessing the relative amount of syngenetic and diagenetic pyrite. *Sedimentary Geology* 308, 18–31.

Plantz, M.R., Fritz, J.S., Smith, F.G. and Houk, R. (1989) Separation of trace metal complexes for analysis of samples of high salt content by inductively coupled plasma mass spectrometry. Analytical Chemistry 61, 149–153.

Qian, G., Brugger, J., Testemale, D., Skinner, W., and Pring, A. (2013) Formation of As(II)-pyrite during experimental replacement of magnetite under hydrothermal conditions. *Geochimica et Cosmochimica Acta* 100, 1–10.

Reed, N.M., Cairns, R.O., Hutton, R.C., and Takaku, Y. (1994) Characterization of polyatomic ion interferences in inductively coupled plasma mass spectrometry using a high resolution mass spectrometer. Journal of analytical atomic spectrometry 9, 881–896.

Reich, M., and Becker, U. (2006) First-principles calculations of the thermodynamic mixing properties of arsenic incorporation into pyrite and marcasite. *Chemical Geology* 225, 278–290.

Rickard, D., and Morse, J. W. (2005) Acid volatile sulfide (AVS). *Marine Chemistry*, 97, 141–197.

Román, N., Reich, M., Leisen, M., Morata, D., Barra, F., and Deditius, A. P. (2019) Geochemical and micro-textural fingerprints of boiling in pyrite. *Geochimica et Cosmochimica Acta* 246, 60–85.

Sahoo, S. K., Planavsky, N. J., Jiang, G., Kendall, B., Owens, J. D., Wang, X., Shi, X., Anbar, A. D., and Lyons, T. W. (2016) Oceanic oxygenation events in the anoxic Ediacaran ocean. *Geobiology*.

Scott, C., Lyons, T. W., Bekker, A., Shen, Y., Poulton, S. W., Chu, X., and Anbar, A. D. (2008) Tracing the stepwise oxygenation of the Proterozoic ocean. *Nature* 452, 456–459.

Steadman, J. A., Large, R. R., Meffre, S., Olin, P. H., Danyushevsky, L. V., Gregory, D. D., Belousov, I., Lounejeva, E., Ireland, T. R., and Holden, P. (2015) Synsedimentary to early diagenetic gold in black shale-hosted pyrite nodules at the Golden Mile Deposit, Kalgoorlie, Western Australia. *Economic Geology* 110, 1157–1191.

Stepanov, A. S., Danyushevsky, L.V., Large, R.R., Mukherjee, I. and Zhukova, I.A. (2020) Deconvolution of the composition of fine-grained pyrite in sedimentary matrix by regression of time-resolved LA-ICP-MS data. American Mineralogist: Journal of Earth and Planetary Materials 105, 820–832.

Swanner, E. D., Webb, S. M., and Kappler, A. (2019) Fate of cobalt and mickel in mackinawite during diagenic pyrite formation. *American Mineralogist* 104, 917–928.

Sykora, S., Cooke, D. R., Meffre, S., Stephanov, A .S., Gardner, K., Scott, R., Selley, D., and Harris, A. C. (2018) Evolution of pyrite trace element compositions from porphyry-style and epithermal conditions at the Lihir gold deposit: implications for ore genesis and mineral processing. *Economic Geology* 113, 193–208.

Tan, S. H., and Horlick, G. (1986) Background spectral features in inductively coupled plasma/mass spectrometry. *Applied Spectroscopy* 40, 445–460.

Tardani, D., Reich, M., Deditius, A. P., Chryssoulis, S., Sánchez-Alfaro, P., Wrage, J.,and Roberts, M. P. (2017) Copper-arsenic decoupling in an active geothermal system: a link between pyrite and fluid composition. *Geochimica et Cosmochimica Acta* 204, 179–204.

Tribovillard, N., Algeo, T. J., Lyons, T., and Riboulleau, A. (2006) Trace metals as paleoredox and paleoproductivity proxies: An update. *Chemical Geology* 232, 12–32.

Vandecasteele, C., Vanhoe, H., and Dams, R. (1993) Inductively coupled plasma mass spectrometry of biological samples. Invited lecture. *Journal of Analytical Atomic Spectrometry* 8, 781–786.

Vorlicek, T. P., Helz, G. R., Chappaz, A., Vue, P., Vezina, A., and Hunter, W. (2018) Molybdenum burial mechanism in sulfidic sediments: iron-sulfide pathway. *ACS Earth and Space Chemistry* 2, 565–576.

Wang, L., Shi, X., and Jiang, G. (2012) Pyrite morphology and redox fluctuations recorded in the Ediacaran Doushantuo Formation. *Palaeogeography, Palaeoclimatology, Palaeoecology* 333, 218–227.

Wacey, D., Kilburn, M. R., Saunders, M., Cliff, J. B., Kong, C., Liu, A. G., Matthews, J. J., and Brasier, M. D. (2015) Uncovering framboidal pyrite biogenicity using nano-scale CNorg mapping. *Geology* 43, 27–30.

Wilkin, R. T., Barnes, H. L., and Brantley, S. L. (1996) The size distribution of framboidal pyrite in modern sediments: An indicator of redox conditions. *Geochimica et Cosmochimica Acta* 60, 3897–3912.

Cambridge Elements ⹀

Elements in Geochemical Tracers in Earth System Science

Timothy Lyons
University of California

Timothy Lyons is a Distinguished Professor of Biogeochemistry in the Department of Earth Sciences at the University of California, Riverside. He is an expert in the use of geochemical tracers for applications in astrobiology, geobiology and Earth history. Professor Lyons leads the 'Alternative Earths' team of the NASA Astrobiology Institute and the Alternative Earths Astrobiology Center at UC Riverside.

Alexandra Turchyn
University of Cambridge

Alexandra Turchyn is a University Reader in Biogeochemistry in the Department of Earth Sciences at the University of Cambridge. Her primary research interests are in isotope geochemistry and the application of geochemistry to interrogate modern and past environments.

Chris Reinhard
Georgia Institute of Technology

Chris Reinhard is an Assistant Professor in the Department of Earth and Atmospheric Sciences at the Georgia Institute of Technology. His research focuses on biogeochemistry and paleoclimatology, and he is an Institutional PI on the 'Alternative Earths' team of the NASA Astrobiology Institute.

About the series

This innovative series provides authoritative, concise overviews of the many novel isotope and elemental systems that can be used as 'proxies' or 'geochemical tracers' to reconstruct past environments over thousands to millions to billions of years—from the evolving chemistry of the atmosphere and oceans to their cause-and-effect relationships with life.

Covering a wide variety of geochemical tracers, the series reviews each method in terms of the geochemical underpinnings, the promises and pitfalls, and the 'state-of-the-art' and future prospects, providing a dynamic reference resource for graduate students, researchers and scientists in geochemistry, astrobiology, paleontology, paleoceanography and paleoclimatology.

The short, timely, broadly accessible papers provide much-needed primers for a wide audience—highlighting the cutting-edge of both new and established proxies as applied to diverse questions about Earth system evolution over wide-ranging time scales.

Cambridge Elements $\equiv$

Elements in Geochemical Tracers in Earth System Science

Elements in the series